The 90-Day Stoic and PMA Challenge

This book was originally published as the First Edition in 2025 and has been revised and updated for this Second Edition.

ISBN 978-82-8485-040-5 (ePub, Second Edition)
ISBN 978-82-8485-039-9 (Softcover, Second Edition)
ISBN 978-82-8485-041-2 (Audiobook, Second Edition)

Second Edition | 2026

Cover design, formatting, and interior layout by PMA Science LLC Publishing.

The 90-Day Stoic and PMA Challenge

A System for Mental Discipline, Emotional Control, and Consistent Action

2nd Edition

Jay Pacheco

CONTENTS

CONTENTS

Quote

Gratitude steadies the mind.
Discipline directs the body.
Action reveals the soul.
—Jay Pacheco

Epigraph

What stands in the way becomes the way.

Not by force, but through discipline, repetition, and acceptance.

The mind is trained.

The body follows.

Dedication

To those who continue—
even when it is difficult.

Author's Note

Why This Book Was Written

In the fall of 2024, my life changed in a way I did not anticipate.

A car accident—caused by another driver's recklessness—brought me to a point where everything I had previously taken for granted required attention, patience, and awareness. Recovery was not only physical; it demanded mental discipline, emotional stability, and a deeper understanding of how I was living.

In the period that followed, I was often asked how I managed to strengthen my mindset, habits, and overall well-being despite the circumstances.

The answer is simple, but not easy.

It began with a decision. What followed would eventually become part of what I now describe as PMA Science and PMA Science Stoic: a practical integration of disciplined thinking, emotional awareness, gratitude, Stoic clarity, and repeated action applied in real life.

A decision to take responsibility for my thoughts, my actions, and the direction of my life. A decision to rely on principles that have endured through time—Stoic philosophy, psychological understanding, and the consistent application of a Positive Mental Attitude (PMA).

What started as a personal commitment gradually became a structured process. Day by day, small actions replaced old patterns. Attention replaced distraction. Discipline replaced inconsistency. Over time, those daily efforts formed what is now the 90-Day Stoic and PMA Challenge.

Since that time, I have continued to apply these principles in my daily life. With distance from the initial experience, the question became not whether the method worked in a moment of adversity, but whether it could sustain clarity, discipline, and direction over time.

It does.

Not because it is easy, but because it is consistent. The process remains effective when it is practiced with honesty and discipline. It does not depend on circumstance, only on application.

This book is a reflection of that process.

It is not written to impress, and it is not intended to offer quick motivation. It provides a framework that must be lived, not observed. The goal is not temporary change, but lasting alignment between thought, action, and purpose.

Real strength is not found in appearances. It is built in the unseen—through consistency, self-awareness, discipline, and the ability to remain steady regardless of external conditions.

The principles in this book are simple, but their impact depends entirely on application.

If you choose to follow them with consistency, the results will follow.

— Jay Pacheco

Prologue

There are moments in life that do not ask for permission. They arrive without warning and quietly divide everything into what came before and what follows after.

In the fall of 2024, I experienced one of those moments.

A car accident disrupted more than just a routine—it forced a complete shift in awareness. Things that once required no thought suddenly demanded attention. Movement became intentional. Rest became necessary. Even the simplest actions had to be approached differently.

There was no immediate sense of clarity, nor any clear understanding of how long recovery would take or what the outcome would be. What remained was the present moment and the responsibility to face it as it was.

2 The 90-Day Stoic and PMA Challenge

At first, the instinct was to resist—to compare the present to the past and question what had changed. But that approach led nowhere. The situation did not respond to frustration or impatience. It required a different response: observation, acceptance, and steady action.

So the focus shifted.

Instead of concentrating on what was no longer possible, attention moved toward what could still be done. Small actions became the starting point. Walking short distances, rebuilding strength gradually, and returning to daily tasks step by step—not to prove anything, but to restore function and stability.

Progress did not happen all at once. It came through repetition.

Driving again required patience. Returning to work required adjustment. Even maintaining energy throughout the day became part of the process. Each step forward, no matter how small, contributed to something larger.

Over time, what began as recovery turned into a form of training—not only for the body, but for the mind.

The body responds to what it is given, but the mind determines what is repeated. And what is repeated, over time, becomes the foundation of how a person lives.

This realization changed the direction entirely.

The situation itself did not need to be controlled. What mattered was the response to it—the ability to remain steady, to act with intention, and to continue forward regardless of external conditions.

That is where this journey began.

Not as an attempt to become someone new, but as a commitment to live with greater awareness, discipline, and clarity than before.

What follows is not a theory. It is a process shaped through experience, tested through repetition, and built on principles that have endured over time.

With time and distance, the experience became clearer. What initially appeared as a disruption revealed itself as direction.

PREFACE

A structure repeated with honesty becomes a way of living.

This book is built on a simple idea: meaningful change is the result of consistent, deliberate action over time.

There is no shortage of information in the modern world. Advice, strategies, and motivation are widely available, yet many remain in the same patterns. The limitation is not knowledge, but the absence of structure and sustained application.

> The 90-Day Stoic and PMA Challenge—a system for training the mind, body, and soul through discipline, gratitude, and action—was created to address this.

It is not a collection of ideas, but a structured process designed to be followed daily. Each step builds on the previous one. Each action reinforces the next. The goal is not intensity, but consistency.

The decision to structure this process over ninety days is intentional. Change requires repetition, and repetition requires time. Short efforts

6 The 90-Day Stoic and PMA Challenge

often produce temporary results. A structured period creates the conditions for stability, discipline, and lasting alignment.

More than a year after its initial creation, this method has been revisited, applied, and tested under different conditions. The purpose was not to confirm whether it worked once, but whether it could be sustained.

It can.

Its effectiveness, however, depends on one principle: loyalty to oneself.

No system works without integrity. No method produces results without consistency. The structure is reliable, but the outcome depends on the individual applying it.

As Epictetus observed, "No man is free who is not master of himself."

And as Marcus Aurelius wrote, "Waste no more time arguing what a good man should be. Be one."

The foundation of this work is what I describe as PMA Science and PMA Science Stoic—a practical system that integrates Stoicism, psychology, Positive Mental Attitude (PMA), gratitude, disciplined thinking, and repeated action into a way of living applicable in real-world conditions.

It is not a theory. It is structure, tested through experience, and sustained through repetition.

Stoicism provides clarity. It focuses attention on what can be controlled, strengthens emotional stability, and encourages disciplined action. It removes unnecessary resistance and replaces it with perspective.

PMA provides direction. It emphasizes constructive thinking, gratitude, and the deliberate choice to approach challenges with intention. Where Stoicism stabilizes, PMA initiates.

Together, they create balance. One grounds. The other moves forward.

Psychological principles support this process by explaining how habits and behaviors are formed. Patterns are shaped through repetition. When repetition is intentional, change becomes predictable.

This book is structured to guide that process in a practical and direct way.

Each week focuses on a specific area of development, progressing from awareness to discipline, emotional control, purpose, and integration. The daily structure is simple, but effective when applied consistently.

There are no guarantees—only principles.

If applied with discipline, they will work. If ignored, they will not.

This is not a book to be read once. It is a process to be lived.

Introduction

What is left unobserved is usually left unchanged.

Awareness, Direction, and the Beginning of Change

Most people do not lack potential. They lack clarity, structure, and consistent application.

The desire to improve is common. The difficulty lies in translating that intention into daily action. Without a clear process, even the strongest motivation fades, and old patterns return.

This is where change often fails—not because it is impossible, but because it is unstructured.

The purpose of this book is to provide structure.

The 90-Day Stoic and PMA Challenge is designed as a practical system that guides you through a process of awareness, discipline, and alignment. It is not based on intensity or short-term effort, but on consistency and deliberate action over time.

10 The 90-Day Stoic and PMA Challenge

The process begins with awareness.

Before anything can be improved, it must first be seen clearly. Thoughts, habits, emotional responses, and daily behaviors operate largely on repetition. Without awareness, they remain unchanged. With awareness, they become adjustable.

From there, discipline is introduced.

Discipline is often misunderstood as restriction. In reality, it is the ability to act in alignment with what is necessary, regardless of mood or circumstance. It creates stability and allows progress to continue even when motivation is absent.

As clarity and discipline deepen, emotional control follows.

Emotions are not obstacles. They are signals. When understood correctly, they provide information rather than interference. Learning to observe them without immediate reaction creates space for better decisions.

Over time, these elements begin to align.

Thoughts become more deliberate. Actions become more consistent. Reactions become more measured. What once required effort becomes part of how you operate.

This is the foundation of lasting change.

The structure of this book reflects that progression.

Each week focuses on a specific area, beginning with awareness and gradually moving toward identity, habits, emotional control, purpose, and integration. The daily challenges are simple by design. Their effectiveness depends on repetition and attention, not complexity.

INTRODUCTION 11

There is no requirement to complete this perfectly.

What matters is participation.

Some days will feel clear and productive. Others will feel resistant. Both are part of the process. Progress is not determined by a single day, but by the willingness to continue.

The responsibility for the outcome remains with you.

This book provides the framework. The results depend on how consistently it is applied.

If approached with honesty and discipline, the process will lead to measurable change—not only in actions, but in how you think, respond, and move through daily life.

This is where the work begins.

THE PRACTICE SYSTEM

HOW TO USE THIS BOOK

Read slowly. Apply deliberately. Record honestly.

This expanded edition is designed to be used, not merely read. The teaching pages establish the principle. The practice pages turn that principle into observation, action, and reflection.

Move through the challenge in order. Complete one day at a time. If a day is missed, continue from where you stopped rather than abandoning the process.

Write briefly but truthfully. A clear sentence grounded in experience is more valuable than a page written only to fill space.

The goal is not perfection. The goal is a visible pattern of greater awareness, steadier discipline, and more deliberate action across ninety days.

THE PRACTICE SYSTEM

THE DAILY METHOD

A repeatable rhythm for each of the ninety days

1. READ

Read the day's principle before the day becomes crowded. Identify the one idea that deserves attention.

2. OBSERVE

Notice the thoughts, emotions, habits, and situations connected to that principle without immediate judgment.

3. APPLY

Choose one concrete action. Keep it small enough to complete and meaningful enough to matter.

4. REFLECT

At the end of the day, record what happened, what became clearer, and what should be repeated or adjusted.

5. CONTINUE

Do not turn one imperfect day into a reason to stop. Consistency is built by returning.

BEFORE DAY 1

STARTING ASSESSMENT

Score the present honestly; this is a baseline, not a judgment.

Circle one score from 1 (low) to 10 (strong).

Awareness of thoughts and reactions	1	2	3	4	5	6	7	8	9	10
Consistency of daily action	1	2	3	4	5	6	7	8	9	10
Emotional steadiness under pressure	1	2	3	4	5	6	7	8	9	10
Clarity of direction and purpose	1	2	3	4	5	6	7	8	9	10
Ability to recover after difficulty	1	2	3	4	5	6	7	8	9	10
Alignment between values and behavior	1	2	3	4	5	6	7	8	9	10

What currently needs the most honest attention?

BEFORE DAY 1

PERSONAL COMMITMENT

Define the direction before beginning the movement.

The change I am choosing to practice over the next ninety days is:

The reason this matters deeply to me is:

The behaviors that will support this commitment are:

When resistance appears, I will return by:

Signature / personal mark

Week 1

Awareness and Foundation

Before the strike, the stillness.
Before the change, the seeing.
What is watched without fear begins to soften.

From the mat:

On October 4, 2024, at 8:35 in the evening, I was driving home through Sandefjord. In the space of a few milliseconds, I heard everything I knew turn—from the ordinary sound of a normal night, to the crash, to a stillness beyond time. The car spun two, three, perhaps four times. Then it stopped.

And in that stillness, before the pain, before the fear, I did one small thing. I put the car into park. I breathed. And I whispered the first words that rose from somewhere deeper than shock: Thank you, Jehovah. I am alive.

That was awareness—the space between what happened and how I answered it. This week, we begin there.

Awareness is the beginning of control.

The beginning of any meaningful change is not action.

It is awareness, decision, gratitude, faith—and then action.

Most people move through their days without truly observing how they think, how they react, or what drives their decisions. Thoughts appear automatic. Emotions feel immediate. Patterns repeat without being questioned—not because change is impossible, but because awareness has not yet been established.

Before anything can be improved, it must first be seen clearly.

This week is not about fixing anything. It is about understanding what is already present.

Awareness is the ability to observe without immediate reaction. It creates distance between what happens and how you respond. And in that distance, there is choice.

That is where change begins.

In Stoic philosophy, attention is placed on what is within one's control—thoughts, actions, and responses. External events are not. This distinction is simple, yet rarely practiced consistently.

The purpose of this week is to begin practicing that distinction in real time:

Observe what happens internally before responding outwardly.

Recognize recurring patterns instead of repeating them unconsciously.

And begin to create space where reaction once existed.

At first, this may feel unfamiliar. The mind is conditioned for speed—for immediate reaction, for noise, for distraction. Slowing down can feel uncomfortable. Observing without acting can feel unnatural.

But this is where control begins.

Not through force—but through awareness.

Day 1 — Observing Without Interference
Begin the day without attempting to change anything.

Instead of controlling your thoughts, observe them. Instead of reacting immediately, pause.

Notice the natural rhythm of your mind—where your attention moves without effort, and which thoughts repeat without invitation.

There is nothing to correct.

Only to see clearly.

Throughout the day, take brief internal steps back. When something happens—whether positive or negative—notice your first reaction before acting on it.

16 THE 90-DAY STOIC AND PMA CHALLENGE

Do not suppress it.
Do not justify it.
Only recognize it.

At the end of the day, write down three simple observations about your thoughts and reactions.

Day 2 — Recognizing Patterns

With awareness established, patterns begin to reveal themselves.

Reactions are rarely random. They follow familiar paths shaped by repetition. The same thoughts arise in similar situations. The same responses follow the same triggers.

Today, observe those patterns.
Notice what creates tension.
Notice which thoughts arise automatically.
Notice which behaviors follow without conscious decision.

Patterns do not change through intention alone. They change through understanding—and through the repetition of new responses.

At the end of the day, write down one pattern you observed clearly.
Describe it without judgment.

Day 3 — Creating Space

Between what happens and how you respond, there is a moment.

Most people pass through that moment unconsciously. Reaction feels immediate—automatic.

Today, slow that moment down.

When something triggers a reaction, pause.

Not for long—just long enough to recognize that you have a choice.

That pause separates impulse from action. It introduces control where there was none.

The mind will resist this. It will try to return to speed and habit.

Continue anyway.

At the end of the day, reflect:

> *Where did you pause, and how did it change your response?*

Day 4 — Accepting What Is Present
Awareness without acceptance creates resistance.

When people begin observing themselves, they often judge what they see. They criticize their thoughts, reactions, or habits. This creates tension—and slows growth.

Today, observe without resistance.

What is present is already there. Ignoring it does not remove it. Resisting it does not change it.

Acceptance is not agreement. It is recognition.

When you accept what is present, you stop wasting energy on internal conflict. That energy becomes available for understanding.

At the end of the day, write down one thing you noticed about yourself that you chose to accept rather than resist.

. . .

Day 5 — Directing Attention
Where attention goes, energy follows.

Without awareness, attention becomes scattered—drawn toward distraction, discomfort, or habit. With awareness, attention can be directed.

Today, practice returning your attention.

Notice when your focus drifts without purpose. Notice when your mind moves toward repetitive or unnecessary thoughts.

Gently bring it back.
Not with force—but with intention.
At the end of the day, reflect:
When did you consciously regain control of your attention?

Day 6 — Clarity Through Reflection
Clarity develops through reflection.

At the end of the day, take time to review your thoughts, reactions, and patterns.

Not to judge—but to understand.

What is becoming clearer?

Where are you still reacting automatically?

Where are you beginning to pause?

Clarity is what remains when observation is consistent, and distraction is reduced.

Write a short reflection capturing what you are beginning to understand about yourself.

Day 7 — Establishing the Foundation

This week is not about transformation. It is about building the foundation.

You have observed your thoughts.
You have recognized patterns.
You have created space.
You have practiced acceptance.
You have directed attention.
You have reflected.

This is the beginning of control.

Not control over everything—but control over how you respond.

And that is where change becomes possible.

Write a final reflection for the week:

> *What have you learned about your patterns?*
> *What is now clearer than before?*
> *What direction do you want to move toward next?*

Weekly Reflection Questions

• What truly surprised me about my thoughts, reactions, or behavior this week?

• Which patterns have I been unconsciously repeating without awareness?

• What emotion most influenced my decisions—and why?

• Where in my life can I apply truth with greater courage and honesty?

Dojo Reflection:

This week you only watched. That is enough. The student learns to see before he learns to move. You have begun to notice the space between what happens and how you answer. Guard that space. It is where your freedom lives. Rest now. Tomorrow, we continue.

WEEK 1

AWARENESS AND FOUNDATION

Daily application pages

DAY 1 **Observing Without Interference**

What did I notice in myself today?

How did I apply this principle in a real situation?

What will I repeat, release, or adjust tomorrow?

DAY 2 **Recognizing Patterns**

What did I notice in myself today?

How did I apply this principle in a real situation?

What will I repeat, release, or adjust tomorrow?

WEEK 1

AWARENESS AND FOUNDATION

Daily application pages

DAY 3 Creating Space

What did I notice in myself today?

How did I apply this principle in a real situation?

What will I repeat, release, or adjust tomorrow?

DAY 4 Accepting What Is Present

What did I notice in myself today?

How did I apply this principle in a real situation?

What will I repeat, release, or adjust tomorrow?

WEEK 1

AWARENESS AND FOUNDATION

Daily application pages

DAY 5 Directing Attention

What did I notice in myself today?

How did I apply this principle in a real situation?

What will I repeat, release, or adjust tomorrow?

DAY 6 Clarity Through Reflection

What did I notice in myself today?

How did I apply this principle in a real situation?

What will I repeat, release, or adjust tomorrow?

WEEK 1

WEEKLY INTEGRATION

Awareness and Foundation

Day 7: Establishing the Foundation

What became clearer this week?

Where did I meet resistance, and how did I respond?

Which action or principle made the greatest difference?

What will I carry deliberately into the next week?

ONE COMMITMENT FOR THE WEEK AHEAD

Week 2

Identity and Discipline

The mountain is not built in a day.
It is built by what returns each day.
Discipline is only love, repeated.

From the mat:

When I was eighteen, I trained in Tae Kwon Do in the Washington, D.C. area, in the school of Grandmaster Jhoon Rhee—the man who first brought the art to America. He would appear at our belt ceremonies, and he told us something simple: instead of watching television, practice your kicks. A hundred of them.

At eighteen, I thought this was almost foolish. At forty-seven, I understand exactly what he meant. Discipline is not a feeling. It is a decision, repeated, until it becomes who you are. By the eightieth kick, the legs burn and the mind begins to beg. You learn not to show it. You learn that identity is built in the repetitions no one sees.

Identity is not found. It is built.

. . .

You do not rise to the level of your goals. You fall to the level of your systems and your self-image.

This week is about defining who you are—and building the discipline required to become that person.

Identity shapes what you believe is possible. Discipline proves it through action.

You do not become who you want to be by chance. You become that person through repeated decisions aligned with who you choose to be.

Discipline is not restriction. It is self-respect in action.

Identity is not something you discover once. It is something you reinforce daily.

Day 8 — Who Am I, Really?
You are not your job.
You are not your past.
You are not what others say about you.
But beneath all of that—who are you?
Today, define yourself with intention.
Write your personal "I am" identity statement:

"I am the kind of person who..."

Define yourself through values, not roles.
Truth. Strength. Discipline. Kindness. Focus. Integrity.
Let it reflect who you are becoming—not who you have been.
Identity, when clearly defined, removes hesitation.

Day 9 — The Discipline Foundation
Discipline is not punishment. It is self-respect expressed through action.

Without discipline, identity remains an idea. With discipline, it becomes reality.

Today, identify three areas in your life where discipline is lacking.
It may be sleep. Nutrition. Focus. Speech. Consistency.
For each area, choose one small, clear action you will repeat daily.
Not something extreme—something sustainable.
Discipline is built through repetition, not intensity.

Day 10 — Ending Internal Negotiation
Every time you break a promise to yourself, your identity weakens.
Every time you follow through, it strengthens.
Most people negotiate with themselves:

"I'll do it later."
"Just this once."
"Tomorrow will be better."

Today, remove negotiation.
Choose one task you know you should do—but usually avoid.
Do it immediately.
No delay. No internal debate.
Action builds self-trust.
Self-trust builds identity.

Day 11 — The Power of Small Wins
Significant change is not built in one moment. It is built through small, repeated victories.
Today, track your wins—no matter how small.
Woke up on time.
Drank water.
Completed a task.
Stayed calm under pressure.

Followed through on a commitment.
These are not minor. They are foundational.
Each small win reinforces your identity.
At the end of the day, review them. Let them accumulate.

Day 12 — Eliminating Distraction
Distraction weakens identity.

Every time you choose distraction over intention, you move further from the person you are becoming.

Today, identify your three main distractions.
It may be your phone. Social media. Unnecessary noise. Avoidance habits.

Remove them for the day.
Create distance between you and what pulls your attention away.
Clarity increases when distraction decreases.

Day 13 — Releasing the Old Identity
Growth requires letting go.
You cannot fully step into a new identity while holding on to the old one.

Today, define what no longer belongs to you.
"I am no longer the person who..."
Complete that sentence honestly.
Say it out loud.
Recognize it. Then release it.
Change begins when you stop identifying with what no longer serves you.

Day 14 — The Discipline Creed
Identity must be reinforced.
Today, create your personal Discipline Creed—a clear statement of who you are and how you act.

Keep it direct. Grounded and true.

Example:
"I am disciplined in my actions.
I follow through on what I commit to.
I act with clarity, not impulse.
I do what is required, regardless of how I feel."
Read it in the morning. Repeat it daily.
Identity strengthens through repetition.

Weekly Reflection Questions

• Who am I becoming through my daily actions—not just my intentions?

• Where did I prove to myself that I can be trusted?

• What small act of discipline had the greatest impact on my mindset?

• Where did I act against my own standards, and why?

Dojo Reflection:

You are not who you say you are. You are what you repeat. This week you practiced the small, unseen acts that no one applauds—and those are the ones that shape a life. When you failed, you saw it without shame and returned. That returning is the discipline. Keep returning.

WEEK 2

IDENTITY AND DISCIPLINE

Daily application pages

DAY 8 **Who Am I, Really?**

What did I notice in myself today?

How did I apply this principle in a real situation?

What will I repeat, release, or adjust tomorrow?

DAY 9 **The Discipline Foundation**

What did I notice in myself today?

How did I apply this principle in a real situation?

What will I repeat, release, or adjust tomorrow?

WEEK 2

IDENTITY AND DISCIPLINE

Daily application pages

DAY 10 **Ending Internal Negotiation**

What did I notice in myself today?

How did I apply this principle in a real situation?

What will I repeat, release, or adjust tomorrow?

DAY 11 **The Power of Small Wins**

What did I notice in myself today?

How did I apply this principle in a real situation?

What will I repeat, release, or adjust tomorrow?

WEEK 2

IDENTITY AND DISCIPLINE

Daily application pages

DAY 12 Eliminating Distraction

What did I notice in myself today?

How did I apply this principle in a real situation?

What will I repeat, release, or adjust tomorrow?

DAY 13 Releasing the Old Identity

What did I notice in myself today?

How did I apply this principle in a real situation?

What will I repeat, release, or adjust tomorrow?

WEEK 2

WEEKLY INTEGRATION

Identity and Discipline

Day 14: The Discipline Creed

What became clearer this week?

Where did I meet resistance, and how did I respond?

Which action or principle made the greatest difference?

What will I carry deliberately into the next week?

ONE COMMITMENT FOR THE WEEK AHEAD

Week 3
Habits and Willpower

Your habits are your future in repeated form.

Your life is shaped by what you do repeatedly.

Not by intention.
Not by motivation.
But by habit.

This week is about building systems that support your identity—and strengthening the inner force that allows you to act consistently: willpower.

Habits create structure.
Willpower initiates action.
Together, they form the foundation of consistency.

You do not need more motivation. You need patterns that remain when motivation fades.

Master your habits, and your direction becomes stable.

Day 15 — The Habit Inventory
You do not rise to your potential. You return to your patterns.

Today, observe your habits with honesty.
Create two columns:
Habits that strengthen you
Habits that weaken you
Do not filter. Write clearly.
Once complete, choose one habit that weakens you.
Remove it—or replace it with a stronger alternative.
Change begins with awareness—but it continues through decision.

Day 16 — The Principle of Small Change
Lasting change does not happen through extremes. It develops through consistency.

Today, choose one habit that improves your life—even slightly.
Keep it small. Keep it clear.
One minute of focus.
One line of writing.
One intentional breath.
One small act of discipline.
What matters is not intensity—but repetition.
Small actions, repeated daily, shape identity over time.

Day 17 — Habit Triggers
Habits are not random. They are responses to cues.

Every habit begins with a trigger.
Today, identify one habit you want to build—and attach it to something you already do.
After waking → take one mindful breath.

28 The 90-Day Stoic and PMA Challenge

After eating → reflect for one minute.
After brushing your teeth → stretch briefly.
This creates continuity.
When actions are connected, they become automatic.

Day 18 — Strengthening Willpower
Willpower is not fixed. It strengthens through use.

Every time you act despite resistance, you build it.
Today, choose one action that requires effort or discomfort.
Wake earlier.
Complete a task you have been avoiding.
Resist an impulse.
Act anyway.
Notice what happens after.
That moment—when resistance is overcome—is where discipline is formed.

Day 19 — The Habit of Thought
Before actions become habits, thoughts do.

Your internal dialogue shapes your external behavior.
Today, observe your thoughts.
When a limiting or negative thought appears, do not ignore it.
Recognize it. Then adjust it.
"I can't do this" becomes
"I am learning how to do this."
"I failed" becomes
"I am improving through this."
Your thoughts guide your direction long before your actions do.

Day 20 — Emotional Patterns
Habits are not only physical. They are emotional.

Some reactions become automatic—stress, frustration, avoidance, doubt.

Today, observe your emotional patterns.
Which emotion appears most often?
When does it arise?
What reinforces it?
Now choose a different response.
Not by force—but by awareness and intention.
Emotions can be redirected when they are recognized early.

A Quiet Adjustment

There were moments when the body reacted before the mind could interpret what was happening.

A sudden discomfort. A shift in energy. A sense that something was not stable.

At first, the response was immediate. Tension followed. Attention narrowed. The reaction came quickly, without space.

Over time, that changed.

Instead of reacting immediately, there was a pause.

Not to suppress the feeling, but to observe it. To understand whether it required action or simply awareness.

Often, it passed.

Not because it was ignored, but because it was not amplified.
This created a different relationship with discomfort.
What once felt urgent became manageable.
And what once controlled the response became something that could be understood.

Day 21 — Systems Over Motivation
Motivation changes. Systems remain.

If you rely on feeling ready, you will act inconsistently.
If you rely on structure, you will act regardless of mood.
Today, design a simple system for one habit.

Define:
When it happens
Where it happens
What triggers it
How you will track it
Make it clear and repeatable.

The goal is not perfection—it is consistency.

Weekly Reflection Questions

• Which habit had the greatest influence on my energy and focus this week?

• Where did I rely on motivation instead of structure?

• What patterns are becoming automatic—both helpful and unhelpful?

• What can I simplify or systematize to reduce decision-making?

WEEK 3

HABITS AND WILLPOWER

Daily application pages

DAY 15 The Habit Inventory

What did I notice in myself today?

How did I apply this principle in a real situation?

What will I repeat, release, or adjust tomorrow?

DAY 16 The Principle of Small Change

What did I notice in myself today?

How did I apply this principle in a real situation?

What will I repeat, release, or adjust tomorrow?

WEEK 3

HABITS AND WILLPOWER

Daily application pages

DAY 17 Habit Triggers

What did I notice in myself today?

How did I apply this principle in a real situation?

What will I repeat, release, or adjust tomorrow?

DAY 18 Strengthening Willpower

What did I notice in myself today?

How did I apply this principle in a real situation?

What will I repeat, release, or adjust tomorrow?

WEEK 3

HABITS AND WILLPOWER

Daily application pages

DAY 19 — The Habit of Thought

What did I notice in myself today?

How did I apply this principle in a real situation?

What will I repeat, release, or adjust tomorrow?

DAY 20 — Emotional Patterns

What did I notice in myself today?

How did I apply this principle in a real situation?

What will I repeat, release, or adjust tomorrow?

WEEK 3

WEEKLY INTEGRATION

Habits and Willpower

Day 21: Systems Over Motivation

What became clearer this week?

Where did I meet resistance, and how did I respond?

Which action or principle made the greatest difference?

What will I carry deliberately into the next week?

ONE COMMITMENT FOR THE WEEK AHEAD

Week 4
Mind and Emotion Mastery

The storm does not ask permission.
But the one who breathes within it
is not carried away.

From the mat:

The day after the accident, I did something I should not have done—I drove. My mind began to slide toward the dark: the pain, the loss, the unfairness of it. I could feel it pulling me down.

So I spoke to my own mind. I said: I am grateful that I am alive. And slowly, my mind began to steady. The pain did not leave—if anything, it grew worse. But gratitude became the thing I held onto. Not to erase the pain, but to keep it from becoming my master. That is emotional mastery: not the absence of the storm, but a hand on the wheel inside it.

Your external life reflects your internal state.

Your thoughts shape your perception.

Your perception influences your emotions.

Your emotions impact your actions.

If the mind is undirected, emotions take control.
If the mind is steady, emotions can be understood—and guided.

This week is about mastering your inner world.
Not by suppressing thoughts or emotions, but by understanding them clearly.

The goal is not to eliminate emotion.
The goal is to experience it without losing direction.

Clarity creates stability.
Stability creates control.

Day 22 — The Mind as a Filter
Your mind does not simply observe reality—it interprets it.

Every thought you repeat becomes a lens through which you see the world.

Today, observe your mental input.

What are you consuming?
What are you repeating internally?
What thoughts are shaping your perception?

Remove what weakens your clarity.
Introduce what strengthens it.
You are responsible for what you allow into your mind.

Day 23 — Meaning and Emotion
Emotions are not created by events alone, but by the meaning you assign to them.

The same situation can produce different emotions depending on interpretation.

Today, observe a situation that triggers a strong emotional response.

Ask:
What meaning am I assigning to this?
Is there another way to interpret it?
When meaning changes, emotional intensity often changes with it.
This is where control begins.

Day 24 — The Observer Within
You are not your thoughts. You are the one aware of them.

When you identify with every thought, you lose clarity.
When you observe them, you gain distance.

Today, practice noticing your thoughts without attaching to them.
Instead of saying, "I am stressed," recognize:

"There is stress present."

This small shift creates space.
And within that space, you regain control.

Day 25 — Naming Emotions Clearly
Clarity reduces confusion.

Many people experience emotions without accurately identifying them.

This leads to unnecessary intensity.
Today, pause several times and ask:

What exactly am I feeling?

Be precise.
Is it frustration—or disappointment?

Is it pressure—or fear?
Is it fatigue—or lack of focus?

The more clearly you define an emotion, the easier it becomes to manage.

Day 26 — The Pause Before Response

Between stimulus and response, there is a moment.

Most reactions happen without awareness of that moment.

Today, extend it.
Before responding—pause.
Take a breath.
Recognize the emotion.
Choose your response.
Even a few seconds can change the outcome.

This is practical, real-time control.

Day 27 — Repetition and Mental Conditioning

The mind adapts through repetition.

What you think repeatedly becomes familiar.
What is familiar becomes automatic.
Today, choose one thought that strengthens your direction.
Keep it simple. Keep it grounded.
Repeat it throughout the day—quietly, consistently.
Over time, repetition influences belief.
Belief influences behavior.

Day 28 — Redirecting Emotional Energy

Emotions carry energy.

If ignored, they build pressure.
If understood, they can be redirected.

Today, take a strong emotion—frustration, tension, restlessness.
Do not suppress it.
Use it.
Channel it into action:

Like Write, Move, Focus, and Create.

Energy does not need to be controlled—it needs to be directed.

Weekly Reflection Questions

- Which emotion influenced me most this week, and how did I respond to it?
- Where did I react automatically instead of responding with awareness?
- What thoughts repeatedly shaped my perception?
- How has my understanding of my internal state changed?

Dojo Reflection:

Emotion is weather. It arrives, it moves, it passes. This week you learned to feel the storm without becoming it—to place a hand on the wheel while the wind still blew. You will not always succeed. No one does. But each time you steady yourself, the storm grows a little smaller, and you grow a little larger.

WEEK 4

MIND AND EMOTION MASTERY

Daily application pages

DAY 22 The Mind as a Filter

What did I notice in myself today?

How did I apply this principle in a real situation?

What will I repeat, release, or adjust tomorrow?

DAY 23 Meaning and Emotion

What did I notice in myself today?

How did I apply this principle in a real situation?

What will I repeat, release, or adjust tomorrow?

WEEK 4

MIND AND EMOTION MASTERY

Daily application pages

DAY 24 — The Observer Within

What did I notice in myself today?

How did I apply this principle in a real situation?

What will I repeat, release, or adjust tomorrow?

DAY 25 — Naming Emotions Clearly

What did I notice in myself today?

How did I apply this principle in a real situation?

What will I repeat, release, or adjust tomorrow?

WEEK 4

MIND AND EMOTION MASTERY

Daily application pages

DAY 26 The Pause Before Response

What did I notice in myself today?

How did I apply this principle in a real situation?

What will I repeat, release, or adjust tomorrow?

DAY 27 Repetition and Mental Conditioning

What did I notice in myself today?

How did I apply this principle in a real situation?

What will I repeat, release, or adjust tomorrow?

WEEK 4

WEEKLY INTEGRATION

Mind and Emotion Mastery

Day 28: Redirecting Emotional Energy

What became clearer this week?

Where did I meet resistance, and how did I respond?

Which action or principle made the greatest difference?

What will I carry deliberately into the next week?

ONE COMMITMENT FOR THE WEEK AHEAD

Week 5
Purpose and Direction

Direction gives weight to action.

Beyond habits, discipline, and emotional control, there is a deeper question:

Why?

Not surface-level goals.
Not external validation.
But direction—something that gives your actions meaning.
Without direction, effort becomes scattered.
With direction, even small actions gain weight.
Purpose is not something distant or abstract.

It is revealed through attention, reflection, and honest observation.

It is often already present—but unnoticed.

This week is about reconnecting with that direction.
Not by force, but by clarity.

Day 29 — What Matters Most
Direction begins with understanding what matters.

Today, step back and reflect.
What do you naturally return to when there is no pressure?
What holds your attention over time?
What feels meaningful, even without reward?
Write your answers simply.
Do not overanalyze.
Notice what is consistent.
Meaning often reveals itself through repetition.

Day 30 — Patterns of Interest
Your attention leaves patterns.

What you return to—again and again—is not random.

Today, look at your past:

> What have you been consistently drawn toward?
> What topics, activities, or ideas continue to appear?
> What have you avoided, even though it feels important?

Patterns of attention often indicate direction.
Clarity increases when you observe without judgment.

Day 31 — Learning from Difficulty
Challenges are not separate from direction—they often point toward it.

What you struggle with can reveal what requires development.

Today, reflect on a past difficulty.
What did it require from you?

What did it teach you?
What did it make clearer?

Difficulty often highlights areas of growth and responsibility.
It is not an obstacle to direction—it is part of it.

Day 32 — Quiet Observation
Clarity does not come from constant input. It develops in stillness.

Today, reduce external noise.

Spend time without distraction—no phone, no media, no unnecessary stimulation.

Observe your thoughts as they settle.

What remains when the noise is removed?
What continues to surface?
Direction becomes clearer when attention is not divided.

Day 33 — Acting in Alignment
Purpose is not found only through thinking. It is revealed through action.

Today, take one small action that feels aligned with what you have been noticing.

Not something extreme.
Something real and immediate.

Write.
Reach out.
Begin something you have delayed.
Action clarifies what thinking alone cannot.

Day 34 — Contribution and Value

Direction is strengthened when it extends beyond yourself.

What you do gains meaning when it contributes to others, to a system, or to something larger.

Today, reflect:
How can your actions provide value?
Where can you contribute—directly or indirectly?

This does not need to be large.
Small, consistent contribution builds meaningful direction over time.

Day 35 — Refining Direction
Clarity develops gradually.

This week is not about final answers. It is about refinement.

Today, review what you have observed:
What feels consistent?
What feels meaningful?
What direction is beginning to take shape?

Write it simply.
Direction does not need to be perfect.
It needs to be clear enough to act on.

A Moment from Experience
There was a period when even the simplest actions required attention.

Eating was no longer automatic. What had once been routine became something to approach carefully. Certain foods created discomfort. Others were not tolerated at all. Over time, it became clear that the body had changed and that it needed to be understood again.

The approach shifted.

Instead of forcing normality, the focus became observation and adjustment. Meals became simpler. Ingredients were chosen deliberately. Skin was removed. Portions were controlled. The goal was not comfort, but stability.

There were moments of frustration. Not because the situation was unclear, but because it required patience.

Progress did not come from intensity. It came from consistency.
Over time, the body responded. Not immediately, but steadily. What once caused discomfort became manageable. What once required effort became routine again.

The lesson was not only physical.

It became clear that the same principle applied to the mind.

When something is not working, forcing it does not create change. Understanding it does.

And when understanding is followed by consistent action, change becomes inevitable.

Weekly Reflection Questions

- What patterns of interest or attention became clearer this week?
- What did I learn from the difficulty that I had not recognized before?
- Where did I act in alignment with what feels meaningful?
- What direction is beginning to form, even if it is not fully defined?

WEEK 5

PURPOSE AND DIRECTION

Daily application pages

DAY 29 What Matters Most

What did I notice in myself today?

How did I apply this principle in a real situation?

What will I repeat, release, or adjust tomorrow?

DAY 30 Patterns of Interest

What did I notice in myself today?

How did I apply this principle in a real situation?

What will I repeat, release, or adjust tomorrow?

WEEK 5

PURPOSE AND DIRECTION

Daily application pages

DAY 31 **Learning from Difficulty**

What did I notice in myself today?

How did I apply this principle in a real situation?

What will I repeat, release, or adjust tomorrow?

DAY 32 **Quiet Observation**

What did I notice in myself today?

How did I apply this principle in a real situation?

What will I repeat, release, or adjust tomorrow?

WEEK 5

PURPOSE AND DIRECTION

Daily application pages

DAY 33 Acting in Alignment

What did I notice in myself today?

How did I apply this principle in a real situation?

What will I repeat, release, or adjust tomorrow?

DAY 34 Contribution and Value

What did I notice in myself today?

How did I apply this principle in a real situation?

What will I repeat, release, or adjust tomorrow?

WEEK 5

WEEKLY INTEGRATION

Purpose and Direction

Day 35: Refining Direction

What became clearer this week?

Where did I meet resistance, and how did I respond?

Which action or principle made the greatest difference?

What will I carry deliberately into the next week?

ONE COMMITMENT FOR THE WEEK AHEAD

DAY 30 CHECKPOINT

MEASURE WHAT HAS CHANGED

Progress becomes clearer when it is observed directly.

Which pattern has weakened since Day 1?

Which constructive behavior is becoming more natural?

Where am I still negotiating with myself?

What evidence shows that I am changing?

DAY 30 CHECKPOINT

RECALIBRATE THE DIRECTION

Keep what works. Adjust what does not. Continue.

The principle I need to practice more deliberately is:

The distraction or behavior I will reduce is:

The action I will repeat consistently during the next phase is:

My renewed commitment, stated in one sentence, is:

Week 6

Positive Mental Attitude and Courage

Courage is not the absence of fear.
It is the will that moves while fear remains.
Even a small step is a victory.

From the mat:

In the hardest days after the accident, something happened that I did not expect. I found that I could smile—even with tears on my face.

This was not pretending. It was not denial. It was the fruit of years of training my mind through Stoicism and a Positive Mental Attitude, long before I ever needed them. When the test came, the training answered. A positive attitude is not the belief that nothing is wrong. It is the courage to face what is wrong and still choose, with your whole will, to rise.

Clarity without action remains incomplete.

. . .

You may understand your thoughts.
You may recognize your patterns.
You may begin to see direction.

But without movement, nothing changes.

This week is about activating what you already know.

Positive Mental Attitude (PMA) is not blind optimism.
It is the disciplined decision to direct your mind toward possibility, even in difficulty.

Courage is not the absence of fear.
It is the ability to act despite it.

Together, they create forward movement.

You do not wait to feel ready.
You act, and readiness follows.

Day 36 — Choosing Your Direction
Your mindset is not fixed. It is chosen—moment by moment.

Today, observe how you respond to challenges.
When something does not go as expected, pause.
Instead of asking, "Why is this happening?"
Ask, "What can be done with this?"
This shift is small—but it changes direction.
Where attention goes, energy follows.

Day 37 — Acting Despite Discomfort
Discomfort is not a signal to stop.

It is often a signal that something matters.

Today, choose one action you have been avoiding.
Not because it is impossible—but because it is uncomfortable.
Take the step.
Make the call.
Speak more clearly.
Begin what you have delayed.
Courage is built through action, not preparation.

Day 38 — Reframing Experience
Events do not define you. Your interpretation of them does.

Today, take something from your past that you labeled as negative.

Look at it again—without resistance.
What did it develop in you?
What did it clarify?
What direction did it influence?
Reframing does not change the past.
It changes how you move forward.

Day 39 — Directing Internal Dialogue
Your internal dialogue influences everything.

Most people speak to themselves without awareness.
Today, listen closely.
When a limiting thought appears, pause.
Do not ignore it.
Do not reinforce it.
Adjust it.
Not with exaggeration—but with accuracy.
Replace distortion with clarity.

The way you speak internally becomes the way you act externally.

Day 40 — Acting with Intention
Confidence does not come before action.

It develops through it.
Today, act as the person you are becoming.
Not perfectly—but intentionally.
How would you speak?
How would you respond?
How would you decide?

Then act accordingly, and identity strengthens when action aligns with intention.

Day 41 — Influencing Your Environment
Your presence affects others—whether you intend it or not.

Today, observe how you influence your environment.
Through your tone.
Your attention.
Your response.
Choose to bring clarity rather than reaction.
Stability instead of tension.
Small interactions shape larger outcomes.

Day 42 — Facing What You Avoid
What you avoid often holds information.

It may be discomfort.
Uncertainty.
Or fear of outcome.
Today, identify one thing you have been avoiding.
Approach it directly.
Not with force—but with clarity.
Action reduces uncertainty.
Avoidance increases it.

Weekly Reflection Questions

- Where did I act despite discomfort, and what changed as a result?

• How did I respond to the challenge—by reacting or by directing my mindset?

• What patterns of avoidance became clear to me?

• How did my internal dialogue influence my actions this week?

Dojo Reflection:

A positive attitude is not pretending the wound is not there. It is choosing, with the wound still open, to rise. This week you learned that you can carry pain and still smile—not to deny it, but to refuse to be ruled by it. That is not weakness dressed as cheer. That is courage. Carry it forward.

WEEK 6

POSITIVE MENTAL ATTITUDE AND COURAGE

Daily application pages

DAY 36 **Choosing Your Direction**

What did I notice in myself today?

How did I apply this principle in a real situation?

What will I repeat, release, or adjust tomorrow?

DAY 37 **Acting Despite Discomfort**

What did I notice in myself today?

How did I apply this principle in a real situation?

What will I repeat, release, or adjust tomorrow?

WEEK 6

POSITIVE MENTAL ATTITUDE AND COURAGE

Daily application pages

DAY 38 **Reframing Experience**

What did I notice in myself today?

How did I apply this principle in a real situation?

What will I repeat, release, or adjust tomorrow?

DAY 39 **Directing Internal Dialogue**

What did I notice in myself today?

How did I apply this principle in a real situation?

What will I repeat, release, or adjust tomorrow?

WEEK 6

POSITIVE MENTAL ATTITUDE AND COURAGE

Daily application pages

DAY 40 **Acting with Intention**

What did I notice in myself today?

How did I apply this principle in a real situation?

What will I repeat, release, or adjust tomorrow?

DAY 41 **Influencing Your Environment**

What did I notice in myself today?

How did I apply this principle in a real situation?

What will I repeat, release, or adjust tomorrow?

WEEK 6

WEEKLY INTEGRATION

Positive Mental Attitude and Courage

Day 42: Facing What You Avoid

What became clearer this week?

Where did I meet resistance, and how did I respond?

Which action or principle made the greatest difference?

What will I carry deliberately into the next week?

ONE COMMITMENT FOR THE WEEK AHEAD

Week 7

Stoicism and Endurance

The reed that bends does not break.
What endures is not what resists,
but what remains.

From the mat:

When I was strong enough, I went back to the gym. I stepped onto the treadmill and began to walk—slowly, carefully. My head felt as though it might explode. I would lose my balance. Every part of me wanted to stop.

But I kept walking. Not because I felt powerful. Because action was the only way forward. I did not conquer the pain. I endured it—one careful step at a time.

Looking back now, I see something I did not see then. Years earlier, in Aikido under Sorrentino Sensei, I had learned to meet a force I could not stop and blend with it instead of breaking against it. I did not think of the mat on that treadmill. But the lesson had been in me all along, waiting. Perhaps that is how training works: it does not always announce itself. It simply becomes part of who you are, and shows up when you need it most.

Endurance is clarity under pressure.

There will always be difficulty.

Circumstances will change.
People will act unpredictably.
Outcomes will not always align with expectations.

What remains within your control is how you respond.

Stoicism is not the absence of feeling.
It is the discipline of remaining steady within it.

Endurance is not passive suffering.
It is the decision to continue—with clarity, without complaint, and with purpose.

This week is about strengthening your ability to remain stable under pressure.

Not by avoiding difficulty.
But by meeting it directly.

Day 43 — What Is Within Your Control
Clarity begins with distinction.

Some things are within your control.

48 THE 90-DAY STOIC AND PMA CHALLENGE

Others are not.
Most frustration comes from confusing the two.
Today, observe your challenges and separate them clearly:

What can you influence?
What must you accept?

Focus your energy only on what is within your control.
Release what is not.

This is the foundation of calm.

Day 44 — Voluntary Discomfort
Comfort weakens resilience when it is constant.

Discomfort, when chosen intentionally, strengthens it.

Today, introduce a small form of voluntary discomfort.
Wake earlier.
Take a cold shower.
Complete something physically or mentally demanding.
Observe your response.
You are capable of more than you assume.

Resilience grows when you move toward difficulty—not away from it.

Day 45 — Following Through
Endurance is built through completion.
Not starting. Not planning. Completing.
Today, choose one task and finish it fully.
No delay. No distraction. No partial effort.
Stay with it until it is done.
Every completed action strengthens discipline.
Every unfinished one weakens it.

Day 46 — Accepting Reality

Resistance creates unnecessary suffering.

What has already happened cannot be changed.
Only your response to it can.
Today, reflect on something you have been resisting.
A situation.
An outcome.
A reality you wish were different.
Acknowledge it fully.
Acceptance does not mean approval.
It means clarity.
And clarity allows movement.

Day 47 — Maintaining Composure
External events do not require immediate reaction.

Composure is a skill.
Today, when tension arises, slow your response.
Pause.
Breathe.
Observe before speaking or acting.
Calm is not weakness.
It is control.
How you respond under pressure defines your stability.

Day 48 — Reframing Difficulty
Difficulty is not always an obstacle.

It is often part of the process.
Today, take a current challenge and examine it differently.
What is it developing in you?
What does it require you to strengthen?
What would happen if you faced it directly?
Perspective does not remove difficulty.
But it changes how you move through it.

Day 49 — Remaining Steady
Strength is not intensity.

It is consistency.
Today, focus on remaining steady—regardless of circumstances.
Do what needs to be done.
Maintain your standards.
Stay aligned with your direction.
Not perfectly—but consistently.
Stability is built through repetition.

Returning to Movement
After the accident, movement carried hesitation.

Driving, in particular, required a different kind of effort. The first attempts were short. A few kilometers. Enough to observe, not to push.

There was awareness of the body. Of tension. Of how the mind responded to speed, to sound, to motion.

Nothing was ignored.

Instead of avoiding the discomfort, it was approached gradually. Repetition replaced hesitation. Distance increased over time. Not quickly, but steadily.

There were other challenges.

Traveling by train introduced a different kind of strain. The movement, the vibration, the noise—each element required adjustment. There were moments when the body resisted it. When the head felt pressure. When rest became necessary.

Still, the process continued.

Not out of force, but out of direction.

Each step was measured. Each experience observed. Each response adjusted.

Over time, what once felt unstable became manageable.

The fear did not disappear all at once. It was replaced, slowly, through action.

And with each repetition, confidence was not assumed—it was built.

Weekly Reflection Questions

- What situation tested my ability to remain steady, and how did I respond?
- Where did I focus on what I could not control instead of what I could?
- What discomfort strengthened me this week?
- How has my response to difficulty changed?

Dojo Reflection:

Endurance is not a single act. It is the willingness to take one more careful step when everything asks you to stop. This week you did not conquer difficulty—you outlasted it. And you learned that the training you once did, long ago, lives in you still, waiting for the day you need it. Trust that. What you build now will meet you later.

WEEK 7

STOICISM AND ENDURANCE

Daily application pages

DAY 43 What Is Within Your Control

What did I notice in myself today?

How did I apply this principle in a real situation?

What will I repeat, release, or adjust tomorrow?

DAY 44 Voluntary Discomfort

What did I notice in myself today?

How did I apply this principle in a real situation?

What will I repeat, release, or adjust tomorrow?

WEEK 7

STOICISM AND ENDURANCE

Daily application pages

DAY 45 **Following Through**

What did I notice in myself today?

How did I apply this principle in a real situation?

What will I repeat, release, or adjust tomorrow?

DAY 46 **Accepting Reality**

What did I notice in myself today?

How did I apply this principle in a real situation?

What will I repeat, release, or adjust tomorrow?

WEEK 7

STOICISM AND ENDURANCE

Daily application pages

DAY 47 **Maintaining Composure**

What did I notice in myself today?

How did I apply this principle in a real situation?

What will I repeat, release, or adjust tomorrow?

DAY 48 **Reframing Difficulty**

What did I notice in myself today?

How did I apply this principle in a real situation?

What will I repeat, release, or adjust tomorrow?

WEEK 7

WEEKLY INTEGRATION

Stoicism and Endurance

Day 49: Remaining Steady

What became clearer this week?

Where did I meet resistance, and how did I respond?

Which action or principle made the greatest difference?

What will I carry deliberately into the next week?

ONE COMMITMENT FOR THE WEEK AHEAD

WEEK 8

SPIRITUAL ALIGNMENT

The river does not force the stone.
It moves around it, and in time, through it.
What is aligned does not strain.

FROM THE MAT:

The obstacles came in waves. Pain that would not relent. Not enough money for medicine. An insurance company that would not reimburse. Child support to carry, and a system that offered no plan. Any one of them could have broken me.

And yet, each time an obstacle rose, so did an unexpected gratitude. When I prayed to Jehovah and met the difficulty with a grateful heart, ideas would come—ways to cope, ways forward, ways to accept what I could not change and move through what I could. The impediment to action became the way. This is alignment: when faith, gratitude, and reason stop being separate tools and begin to move as one.

After discipline, action, and endurance, there is another level of clarity:

. . .

Alignment.

Not everything is solved through effort.
Not everything requires force.

There are moments where direction becomes clearer not by doing more —but by becoming still enough to see.

Spiritual alignment is not about belief systems or labels.
It is about coherence—between thought, action, and inner direction.

When you are aligned, decisions become simpler.
Energy becomes more stable.
Action becomes more natural.

This week is about reducing noise—and strengthening that alignment.

Day 50 — Creating Space
Clarity requires space.

If your attention is constantly occupied, there is no room for deeper understanding.

Today, create intentional space.
Reduce unnecessary input.
Limit distractions.
Allow moments of silence.
Notice what arises when there is less noise.
What remains often matters most.

Day 51 — Listening Without Interference
Most people listen to respond—not to understand.

This applies not only to others—but to themselves.

Today, practice listening without interruption.
When a thought arises, observe it without immediately analyzing or reacting.

When someone speaks, listen fully before forming a response.
Understanding deepens when interference is reduced.

Day 52 — Letting Go of Control
Not everything can be controlled.

Trying to control what is uncertain creates tension.
Today, identify one area where you are forcing an outcome.
Step back.
Focus only on what you can influence.
Release what you cannot.
Letting go is not giving up.
It is redirecting energy toward what is within your reach.

Day 53 — Presence in Action
Alignment is not separate from daily life.

It is expressed through it.
Today, bring full attention to simple actions.
Walking, Eating, Working, and Speaking.
Do each without rushing.
Presence is not about slowing everything down—it is about being fully where you are.
When attention is undivided, clarity increases.

Day 54 — Internal Alignment Check
Misalignment often appears as tension, confusion, or fatigue.

Today, observe yourself honestly.
Are your actions aligned with your values?

Are your decisions aligned with your direction?
Are your thoughts aligned with what you know to be true?
Where there is misalignment, there is friction.
Clarity begins by recognizing it.

Day 55 — Trusting the Process
Not every step needs to be visible.

There are moments where progress is not obvious—but it is still occurring.
Today, practice patience.
Continue your actions without needing immediate confirmation.
Trust develops through consistency—not certainty.

Day 56 — Returning to Stillness
You can always return to clarity.

Even when distracted.
Even when uncertain.
Today, take time to sit in stillness.
No input. No output.
Just observation.

Let your thoughts settle without interference.

Clarity is not created—it is revealed when noise is reduced.

Weekly Reflection Questions

- When did I feel most clear or aligned this week?

- What sources of noise or distraction affected my clarity?

- Where did I try to control what was not within my control?

- How has stillness influenced my thinking and decisions?

. . .

Dojo Reflection:

There is effort, and there is alignment. This week you learned that not everything is won by force—that when faith, gratitude, and reason move as one, the way forward often appears on its own. When the next obstacle rises, meet it not with strain, but with alignment. Accept what you cannot change. Move through what you can. And be grateful in both.

WEEK 8

SPIRITUAL ALIGNMENT

Daily application pages

DAY 50 — Creating Space

What did I notice in myself today?

How did I apply this principle in a real situation?

What will I repeat, release, or adjust tomorrow?

DAY 51 — Listening Without Interference

What did I notice in myself today?

How did I apply this principle in a real situation?

What will I repeat, release, or adjust tomorrow?

WEEK 8

SPIRITUAL ALIGNMENT

Daily application pages

DAY 52 Letting Go of Control

What did I notice in myself today?

How did I apply this principle in a real situation?

What will I repeat, release, or adjust tomorrow?

DAY 53 Presence in Action

What did I notice in myself today?

How did I apply this principle in a real situation?

What will I repeat, release, or adjust tomorrow?

WEEK 8

SPIRITUAL ALIGNMENT

Daily application pages

DAY 54 **Internal Alignment Check**

What did I notice in myself today?

How did I apply this principle in a real situation?

What will I repeat, release, or adjust tomorrow?

DAY 55 **Trusting the Process**

What did I notice in myself today?

How did I apply this principle in a real situation?

What will I repeat, release, or adjust tomorrow?

WEEK 8

WEEKLY INTEGRATION

Spiritual Alignment

Day 56: Returning to Stillness

What became clearer this week?

Where did I meet resistance, and how did I respond?

Which action or principle made the greatest difference?

What will I carry deliberately into the next week?

ONE COMMITMENT FOR THE WEEK AHEAD

Week 9

Inner Strength and Presence

Strength is consistency, not intensity.

STRENGTH IS OFTEN MISUNDERSTOOD.

It is not aggression.
It is not force.
It is not constant intensity.

True strength is quiet.
It is stable.
It is controlled.

It does not react impulsively.
It responds with clarity.

This week is about developing that kind of strength—the kind that does not depend on external conditions.

58 THE 90-DAY STOIC AND PMA CHALLENGE

Presence is part of that strength.

When your attention is steady, your actions become precise.
When your mind is clear, your decisions become direct.

This is where control becomes visible.

Day 57 — Defining Your Standard
Strength begins with clarity.

If you do not define your standards, your environment will define them for you.

Today, decide:

What do you stand for?
What do you tolerate?
What do you no longer accept?

Write your standards clearly.
Standards guide behavior when emotions fluctuate.

Day 58 — Strength Through Preparation
Strength is built before it is needed.

When situations become difficult, you rely on what you have practiced —not what you intend.

Today, prepare intentionally.
Strengthen your body.
Focus your mind.
Reinforce your habits.
Preparation reduces uncertainty.
And reduces hesitation.

Day 59 — Facing Internal Resistance

The greatest resistance is often internal.

Delay. Doubt. Avoidance.
These do not come from external pressure—but from internal conflict.

Today, observe where you hesitate.
Then act anyway.
Not perfectly—but directly.
Strength develops when action overrides hesitation.

A Shift in Direction
There was a point where doing more was no longer the solution.

Effort was present. Action was consistent. Yet the results did not align.

That required a different approach.
Instead of increasing intensity, the focus shifted to clarity.
What was actually necessary?
What was being repeated without purpose?
What could be removed?

Some actions were reduced. Others were adjusted. Attention became more selective.

Progress followed.

Not because more was done, but because what remained was aligned.
It became clear that direction matters more than volume.
And that clarity often requires removing what is unnecessary.

Day 60 — Responding Under Pressure
Pressure reveals patterns.

60 THE 90-DAY STOIC AND PMA CHALLENGE

It shows how you think, how you react, and where you lose control.
Today, observe yourself in moments of pressure.
When something unexpected happens:
Pause.
Stabilize.
Respond deliberately.
Do not rush to react.
Control is measured in how you respond when it matters.

Day 61 — Focused Presence
Scattered attention weakens performance.

Focused attention strengthens it.
Today, do one task at a time—with full attention.
No multitasking.
No divided focus.
When your attention is fully present, your output improves—and your mind becomes more stable.

Day 62 — Strength Without Reaction
Not every situation requires a response.

Reacting to everything drains energy.
Today, observe where you react unnecessarily.
A comment.
A delay.
An inconvenience.
Pause.
Ask yourself:
Is a response required?
Often, it is not.
Strength includes restraint.

Day 63 — Recovery and Stability
Strength is not only built through action, but also through recovery.

Without recovery, performance declines.
Today, focus on restoring your energy.
Rest intentionally.
Reflect briefly.
Allow your system to reset.
Stability is maintained through balance.

Weekly Reflection Questions

- Where did I demonstrate control under pressure?
- What internal resistance did I face, and how did I respond?
- When did I act with clarity instead of reacting impulsively?
- How has my definition of strength changed?

WEEK 9

INNER STRENGTH AND PRESENCE

Daily application pages

DAY 57 Defining Your Standard

What did I notice in myself today?

How did I apply this principle in a real situation?

What will I repeat, release, or adjust tomorrow?

DAY 58 Strength Through Preparation

What did I notice in myself today?

How did I apply this principle in a real situation?

What will I repeat, release, or adjust tomorrow?

WEEK 9

INNER STRENGTH AND PRESENCE

Daily application pages

DAY 59 Facing Internal Resistance

What did I notice in myself today?

How did I apply this principle in a real situation?

What will I repeat, release, or adjust tomorrow?

DAY 60 Responding Under Pressure

What did I notice in myself today?

How did I apply this principle in a real situation?

What will I repeat, release, or adjust tomorrow?

WEEK 9

INNER STRENGTH AND PRESENCE

Daily application pages

DAY 61 — Focused Presence

What did I notice in myself today?

How did I apply this principle in a real situation?

What will I repeat, release, or adjust tomorrow?

DAY 62 — Strength Without Reaction

What did I notice in myself today?

How did I apply this principle in a real situation?

What will I repeat, release, or adjust tomorrow?

WEEK 9

WEEKLY INTEGRATION

Inner Strength and Presence

Day 63: Recovery and Stability

What became clearer this week?

Where did I meet resistance, and how did I respond?

Which action or principle made the greatest difference?

What will I carry deliberately into the next week?

ONE COMMITMENT FOR THE WEEK AHEAD

DAY 60 CHECKPOINT

MEASURE WHAT HAS CHANGED

Progress becomes clearer when it is observed directly.

Which pattern has weakened since Day 1?

Which constructive behavior is becoming more natural?

Where am I still negotiating with myself?

What evidence shows that I am changing?

DAY 60 CHECKPOINT

RECALIBRATE THE DIRECTION

Keep what works. Adjust what does not. Continue.

The principle I need to practice more deliberately is:

The distraction or behavior I will reduce is:

The action I will repeat consistently during the next phase is:

My renewed commitment, stated in one sentence, is:

Week 10

Wisdom and Integration

Knowledge alone does not create change.

Understanding does not guarantee action.
Insight does not ensure consistency.

Wisdom is different.

Wisdom is applied understanding.

It is the ability to act clearly, consistently, and with intention—across different situations.

It is not built in a single moment.
It develops through reflection, correction, and repetition.

This week is about integrating what you have learned.

Not as separate ideas—but as a unified way of thinking and acting.

Day 64 — From Understanding to Action

It is possible to understand something—and still not live it.

Today, observe where this happens.
What have you learned that you are not yet applying?
What do you know—but still avoid?
Choose one insight.
Act on it directly.
Wisdom begins when understanding becomes behavior.

Day 65 — Slowing Down to See Clearly

Speed can create errors.

When everything moves quickly, awareness decreases.
Today, slow your pace slightly.
Not to delay—but to observe more clearly.
Think before responding.
Act with intention.
Allow space between decisions.
Clarity increases when speed is reduced.

Day 66 — Honest Self-Assessment

Growth requires accuracy.

Without honest observation, patterns remain unchanged.
Today, reflect on yourself directly:

> *Where are you still reacting automatically?*
> *Where are you inconsistent?*
> *Where are you avoiding responsibility?*

Do not judge—observe.
Accuracy leads to adjustment.
Adjustment leads to improvement.

Day 67 — Remaining Open to Learning

Confidence without openness leads to stagnation.

Wisdom requires the ability to adjust when necessary.
Today, listen carefully.
To others.
To feedback.
To your own experience.
Do not defend immediately.
Do not reject quickly.
Consider what is useful.
Learning continues when resistance is reduced.

Day 68 — Listening with Attention
Most communication is reactive.

People listen to respond—not to understand.
Today, practice listening fully.
Do not interrupt.
Do not prepare your reply while the other person is speaking.
Focus completely on what is being said.
A clear understanding reduces unnecessary conflict.

Day 69 — Acting with Consistency
Consistency builds trust—both internally and externally.

Today, focus on aligning your actions with your standards.
Do what you say you will do.
Follow through on commitments.
Maintain your direction.
Not occasionally—but consistently.
Consistency turns intention into identity.

Day 70 — Applying What You Know
Knowledge that is not used loses value.

Today, take something you have learned over the past weeks—and apply it in a real situation.
Not in theory. In action.
Observe the result.
Adjustment will follow.
Wisdom grows through application, not accumulation.

Weekly Reflection Questions

• Where did I apply what I already knew, instead of seeking more information?

• What did I recognize about myself through honest observation?

• Where did I remain open to learning instead of reacting defensively?

• How has my behavior changed as a result of what I've practiced?

WEEK 10

WISDOM AND INTEGRATION

Daily application pages

DAY 64 From Understanding to Action

What did I notice in myself today?

How did I apply this principle in a real situation?

What will I repeat, release, or adjust tomorrow?

DAY 65 Slowing Down to See Clearly

What did I notice in myself today?

How did I apply this principle in a real situation?

What will I repeat, release, or adjust tomorrow?

WEEK 10

WISDOM AND INTEGRATION

Daily application pages

DAY 66 **Honest Self-Assessment**

What did I notice in myself today?

How did I apply this principle in a real situation?

What will I repeat, release, or adjust tomorrow?

DAY 67 **Remaining Open to Learning**

What did I notice in myself today?

How did I apply this principle in a real situation?

What will I repeat, release, or adjust tomorrow?

WEEK 10

WISDOM AND INTEGRATION

Daily application pages

DAY 68 **Listening with Attention**

What did I notice in myself today?

How did I apply this principle in a real situation?

What will I repeat, release, or adjust tomorrow?

DAY 69 **Acting with Consistency**

What did I notice in myself today?

How did I apply this principle in a real situation?

What will I repeat, release, or adjust tomorrow?

WEEK 10

WEEKLY INTEGRATION

Wisdom and Integration

Day 70: Applying What You Know

What became clearer this week?

Where did I meet resistance, and how did I respond?

Which action or principle made the greatest difference?

What will I carry deliberately into the next week?

ONE COMMITMENT FOR THE WEEK AHEAD

Week 11

Self-Leadership

Self-leadership begins where excuses end.

All change begins with self-direction.

Before you lead anything externally, you must be able to lead yourself.

Your thoughts.
Your actions.
Your decisions.
Your time.

Without self-leadership, progress depends on circumstances.
With it, progress becomes consistent.

Self-leadership is not control in the rigid sense.

It is the alignment between what you know, what you decide, and what you do.

This week is about taking full responsibility for that alignment.

Day 71 — Defining Your Direction
Leadership requires clarity.
If your direction is unclear, your actions will be inconsistent.
Today, define your direction in simple terms:
What are you building?
What do you stand for?
What do your actions represent?
Keep it clear. Keep it grounded.
Direction does not need to be complex—it needs to be consistent.

Day 72 — Following Through Without Delay
Self-leadership is expressed through action.
Not intention. Not planning. Action.
Today, choose one task that matters—and complete it without delay.
No negotiation.
No postponement.
Start. Continue. Finish.
Every completed action strengthens your ability to lead yourself.

Day 73 — Leading Your Emotional State
Your emotional state influences your decisions.
If unmanaged, it creates inconsistency.
Today, observe your emotional state throughout the day.
When it shifts, pause.

Ask:
Is this state supporting my direction?

If not—adjust.
Not by forcing emotion, but by redirecting attention and action.
Leadership includes managing your internal state.

Continuing Without Certainty

There were periods where the outcome was unclear.

Progress was not always visible. The direction was steady, but the result could not be measured immediately.

This created uncertainty.

Not about what to do, but about how long it would take.
The response was simple.
Continue.

Not with expectation, but with commitment to the process.
Some days felt stable. Others required adjustment. But the structure remained.

Over time, results appeared.
Not suddenly, but gradually.

This reinforced a principle:
You do not need certainty to continue.
You need consistency.

And when consistency is maintained, clarity follows.

Day 74 — Acting as an Example

Your actions influence more than yourself.
Even without intention, others observe behavior.
Today, act in a way that reflects your standards.
Speak clearly.
Follow through.
Remain composed.

Do not act for recognition—but for alignment.
Leadership is demonstrated, not declared.

Day 75 — Taking Full Responsibility
Blame delays progress.
Responsibility creates movement.
Today, look at one area of your life where results are not as you want them to be.

Health. Work. Relationships. Direction.

Ask:
What is within my control here?
What have I avoided?
What can I do next?

Focus on action—not explanation.
Responsibility restores control.

Day 76 — Evaluating Your Environment
Your environment influences your behavior.
It can support your direction—or weaken it.

Today, observe your environment:
Who do you spend time with?
What do you consume?
What patterns are reinforced?

Make adjustments where needed.

Alignment becomes easier when your environment supports it.

Day 77 — Leading with Stability
Leadership is not intensity—it is consistency.
Today, focus on maintaining stability.
Follow your structure.

70 THE 90-DAY STOIC AND PMA CHALLENGE

Maintain your standards.
Respond with clarity.
Not occasionally—but throughout the day.
Stability builds trust—in yourself and in your direction.

Weekly Reflection Questions

- Where did I lead myself clearly—and where did I hesitate?
- What decisions strengthened my direction this week?
- Where did I take responsibility instead of placing blame?
- How has my ability to direct myself improved?

WEEK 11

SELF-LEADERSHIP

Daily application pages

DAY 71 — Defining Your Direction

What did I notice in myself today?

How did I apply this principle in a real situation?

What will I repeat, release, or adjust tomorrow?

DAY 72 — Following Through Without Delay

What did I notice in myself today?

How did I apply this principle in a real situation?

What will I repeat, release, or adjust tomorrow?

WEEK 11

SELF-LEADERSHIP

Daily application pages

DAY 73 Leading Your Emotional State

What did I notice in myself today?

How did I apply this principle in a real situation?

What will I repeat, release, or adjust tomorrow?

DAY 74 Acting as an Example

What did I notice in myself today?

How did I apply this principle in a real situation?

What will I repeat, release, or adjust tomorrow?

WEEK 11

SELF-LEADERSHIP

Daily application pages

DAY 75 **Taking Full Responsibility**

What did I notice in myself today?

How did I apply this principle in a real situation?

What will I repeat, release, or adjust tomorrow?

DAY 76 **Evaluating Your Environment**

What did I notice in myself today?

How did I apply this principle in a real situation?

What will I repeat, release, or adjust tomorrow?

WEEK 11

WEEKLY INTEGRATION

Self-Leadership

Day 77: Leading with Stability

What became clearer this week?

Where did I meet resistance, and how did I respond?

Which action or principle made the greatest difference?

What will I carry deliberately into the next week?

ONE COMMITMENT FOR THE WEEK AHEAD

Week 12

Renewal and Reinforcement

The old leaf must fall
for the new to open.
Renewal is not gain, but release.

From the mat:

The accident changed my body in ways that remain with me. My metabolism shifted entirely. I had to become a researcher of my own health —learning what my body could now accept. Today I eat simply: salmon, chicken, turkey, root vegetables, a few honest things. I make a beetroot juice that has become part of my daily practice. I still carry chronic pain in my neck and right hip, and from time to time I visit the chiropractor.

I tell you this so you know: what is in this book is not theory. At forty-seven, I still test it on myself, every day, sometimes through pain. Renewal is not a finish line. It is a way of living—rebuilding, adjusting, continuing. The training never ends. It simply becomes your life.

Change is not only about building new patterns.

. . .

It is also about releasing what no longer serves you.

If old patterns remain active, they will continue to influence your behavior—often without awareness.

Renewal requires both:
Letting go.
And reinforcing what remains.

This week is about strengthening what you have built—and removing what interferes with it.

Not through force.
Through clarity and deliberate action.

Day 78 — Letting Go of What No Longer Serves You
You cannot fully move forward while holding on to patterns that belong to the past.

Today, identify what you need to release.

A habit, a belief, a repeated reaction, and an attachment to a previous version of yourself.

Write it clearly:
"This no longer serves me."

Recognition is the first step toward separation.

Day 79 — Understanding What Has Changed
Progress is not always visible in a single moment—but it becomes clear over time.

Today, reflect on how you have changed.

How do you think differently?
How do you respond differently?
What patterns have weakened?

Write your observations.
Clarity reinforces progress.

Day 80 — Reinforcing Your Identity
What you repeat becomes stable.
Today, reinforce the identity you have been building.
Act in alignment with it—deliberately.
Follow through on your habits.
Maintain your standards.
Respond with awareness.
Identity strengthens through consistent expression.

Day 81 — Reframing the Past
The past does not need to define you.
But how you interpret it will influence your direction.
Today, revisit a past experience that once shaped your behavior.
Observe it without distortion.

What did it teach you?
How has it influenced your decisions?
What can you take from it—without carrying it forward?

Reframing creates separation from old patterns.

Day 82 — Returning to Curiosity
Growth is supported by curiosity.
Without it, thinking becomes fixed.
Today, approach your day with openness.
Ask questions.
Observe without assumption.

Notice details you would normally ignore.
Curiosity keeps your thinking flexible—and your awareness active.

Day 83 — Strengthening Through Contribution
What you have built is not only for yourself.
It can be applied beyond you.
Today, contribute in a simple, direct way.
Offer clarity.
Support someone.
Share something useful.
Contribution reinforces understanding.

Day 84 — Stabilizing the Process
Consistency turns progress into stability.

Today, focus on maintaining what you have built.
Not by adding more—but by sustaining what is already working.
Repeat your habits.
Maintain your awareness.
Follow your structure.
Stability is built through repetition, not intensity.

Weekly Reflection Questions

- What patterns or behaviors did I clearly recognize as no longer useful?

- How have my thinking and response changed since the beginning?

- Where did I reinforce the identity I am building?

- What am I now maintaining consistently that I once struggled with?

Dojo Reflection:

You are near the end of the structured path—but the training does not end. It becomes your life. This week you learned that renewal is not a finish line but a way of living: to release what no longer serves you, to rebuild what does, and to continue, again and again, sometimes through pain. This is the discipline that outlasts the ninety days. This is the discipline of a lifetime.

WEEK 12

RENEWAL AND REINFORCEMENT

Daily application pages

DAY 78 Letting Go of What No Longer Serves You

What did I notice in myself today?

How did I apply this principle in a real situation?

What will I repeat, release, or adjust tomorrow?

DAY 79 Understanding What Has Changed

What did I notice in myself today?

How did I apply this principle in a real situation?

What will I repeat, release, or adjust tomorrow?

WEEK 12

RENEWAL AND REINFORCEMENT

Daily application pages

DAY 80 **Reinforcing Your Identity**

What did I notice in myself today?

How did I apply this principle in a real situation?

What will I repeat, release, or adjust tomorrow?

DAY 81 **Reframing the Past**

What did I notice in myself today?

How did I apply this principle in a real situation?

What will I repeat, release, or adjust tomorrow?

WEEK 12

RENEWAL AND REINFORCEMENT

Daily application pages

DAY 82 **Returning to Curiosity**

What did I notice in myself today?

How did I apply this principle in a real situation?

What will I repeat, release, or adjust tomorrow?

DAY 83 **Strengthening Through Contribution**

What did I notice in myself today?

How did I apply this principle in a real situation?

What will I repeat, release, or adjust tomorrow?

WEEK 12

WEEKLY INTEGRATION

Renewal and Reinforcement

Day 84: Stabilizing the Process

What became clearer this week?

Where did I meet resistance, and how did I respond?

Which action or principle made the greatest difference?

What will I carry deliberately into the next week?

ONE COMMITMENT FOR THE WEEK AHEAD

Week 13

Integration and Continuation

What you can return to, you can rebuild.

This is not a finish line.

It is a transition.

Over the past weeks, you have developed awareness, discipline, emotional control, clarity, and direction.

Now the focus shifts:
From practice → to integration.
From structure → to ownership.
From effort → to consistency.

What you have built must now function without constant guidance.

This week is about confirming that.

Day 85 — Returning to Awareness
Everything began with awareness.
Without it, patterns return unnoticed.
Today, revisit that foundation.
Observe your thoughts.
Observe your reactions.
Notice what has changed—and what has not.
Clear observation remains the starting point, no matter how far you progress

Day 86 — Recognizing Remaining Resistance
Even after progress, resistance can remain.
It may appear as hesitation, avoidance, or subtle patterns.
Today, observe where resistance still exists.

Where do you delay?
Where do you avoid clarity?
Where do old reactions still appear?

Do not ignore it.
Recognize it directly.

Day 87 — Protecting Your Direction
Not everything deserves your attention.
Your focus determines your progress.
Today, evaluate what you allow into your environment:
Information.
Conversations.
Habits.
Inputs.
Remove what disrupts your clarity.
Protect what supports your direction.

Day 88 — Reviewing Your Process
Integration requires reflection.
Today, review your process from the beginning.

What habits have you built?
What patterns have changed?
What principles have become consistent?

Write a clear summary.
Not to evaluate—but to understand.

Day 89 — Applying Without Structure
Until now, structure has guided your actions.
Now, begin acting without relying on it.

Today, move through your day using what you have learned—without prompts.

Make decisions based on awareness.
Respond with clarity.
Maintain your direction.
This is where practice becomes natural.

Day 90 — Continuing Forward
This is not the conclusion of the process.
It is the point where responsibility becomes fully yours.
There will be no perfect consistency.
There will be moments of distraction or misalignment.
That is expected.

What matters is your ability to return to awareness, to structure, to clarity.

Today, define how you will continue.
What will you maintain?
What will you refine?
What will you build next?
Write it clearly.
Then continue.

Weekly Reflection Questions

- What has become consistent without effort?
- Where do I still need awareness and adjustment?
- What principles will I continue to apply moving forward?
- How will I maintain direction beyond this structure?

WEEK 13

INTEGRATION AND CONTINUATION

Daily application pages

DAY 85 Returning to Awareness

What did I notice in myself today?

How did I apply this principle in a real situation?

What will I repeat, release, or adjust tomorrow?

DAY 86 Recognizing Remaining Resistance

What did I notice in myself today?

How did I apply this principle in a real situation?

What will I repeat, release, or adjust tomorrow?

WEEK 13

INTEGRATION AND CONTINUATION

Daily application page

DAY 87 **Protecting Your Direction**

What did I notice in myself today?

How did I apply this principle in a real situation?

What will I repeat, release, or adjust tomorrow?

WEEK 13

INTEGRATION AND CONTINUATION

Daily application page

DAY 88 Reviewing Your Process

What did I notice in myself today?

How did I apply this principle in a real situation?

What will I repeat, release, or adjust tomorrow?

WEEK 13

WEEKLY INTEGRATION

Integration and Continuation

Day 89: Applying Without Structure

What became clearer this week?

Where did I meet resistance, and how did I respond?

Which action or principle made the greatest difference?

What will I carry deliberately into the next week?

ONE COMMITMENT FOR THE WEEK AHEAD

Day 90

The Continuation

The structure ends. The practice begins.

This is not the end. It is a moment of recognition—a quiet pause in which you step back and see with greater clarity what has been built within you.

Ninety days ago, you began with intention. Not certainty, not perfection, but a decision to observe, to act, and to continue. At the time, the path may not have been fully clear. You moved forward anyway.

Now, something has changed.

Not in a sudden or dramatic way, but through repetition—through the small, consistent moments that often go unnoticed while they are happening. You have learned to watch your thoughts rather than be carried by them. You have felt your reactions without immediately becoming them. You have taken actions, even when they felt difficult, even when progress seemed slow.

You did not transform overnight. You built, step by step.

There were days when focus was strong, and direction was clear. There were others where distraction returned, where the mind wandered, where the effort felt heavier than expected. Yet you continued. Not perfectly, but consistently. And consistency, over time, becomes structure.

This is where the real shift has taken place.

Because what you have developed is not temporary discipline, but awareness. And awareness changes everything. It allows you to recognize when you are drifting, to notice when old patterns attempt to return, and to pause before repeating them. Where there was once an automatic reaction, there is now space. And within that space, there is choice.

This is not control over life itself. It is something more grounded, more reliable. It is control over how you meet what happens.

You now understand that direction does not come from intensity, but from repetition. That clarity is not something found once, but something maintained. That strength is not the absence of difficulty, but the ability to remain steady within it.

The structure you followed over these ninety days was never meant to confine you. It was meant to train you. And now, the structure begins to dissolve—not because it is no longer needed, but because it has become part of how you think and act.

What remains is simple, but not easy: to continue.

To observe when it would be easier not to.

To reflect when distraction is available.

To adjust when resistance appears.

To act when hesitation arises.

And then to repeat, without needing external pressure—only internal alignment.

This is no longer a challenge. It is a way of living.

So today is not about finishing. It is about choosing, once again, how you will continue.

Not according to who you were before these ninety days, but according to what you now understand.

Take a moment and be honest with yourself.

What has strengthened that you will continue to build?

What can be refined with greater attention and intention?

And what no longer belongs—what can now be released without resistance?

Write it clearly. Not as a distant goal, but as a present direction.

. . .

Then continue.

Quietly. Deliberately. Without needing to prove anything.

Because what you practice becomes how you live.

And how you live, over time, becomes who you are.

Final Reflection

In what ways has your thinking become more deliberate and your responses more controlled?

Which patterns have lost their influence, no longer directing your behavior without awareness?

What habits or structures have become part of how you operate—quietly, consistently, without force?

And now, with greater clarity than before, what direction are you choosing—not for a moment, but as a way forward?

AFTER DAY 90

FINAL ASSESSMENT

Return to the same dimensions you scored before Day 1.

Awareness of thoughts and reactions	1 2 3 4 5 6 7 8 9 10
Consistency of daily action	1 2 3 4 5 6 7 8 9 10
Emotional steadiness under pressure	1 2 3 4 5 6 7 8 9 10
Clarity of direction and purpose	1 2 3 4 5 6 7 8 9 10
Ability to recover after difficulty	1 2 3 4 5 6 7 8 9 10
Alignment between values and behavior	1 2 3 4 5 6 7 8 9 10

Where is the clearest evidence of change?

AFTER DAY 90

EVIDENCE OF CHANGE

Name what is different without exaggeration and without minimizing it.

I now respond differently when:

The habit that has become more stable is:

The belief or pattern I no longer need is:

The quality I trust more deeply in myself is:

CONTINUATION

THE NEXT NINETY DAYS

A finished challenge should become a living practice.

Three practices I will continue:

One boundary I will protect:

One relationship or responsibility I will approach differently:

The direction I choose for the next ninety days:

CONTINUATION

A PERSONAL CREED

Write the principles you are prepared to live when no one is watching.

Signed Date

Bonus

From Structure to Living

Discipline is not what you do once. It is how you live.

What you have completed was structured.

Each day had a direction. Each step had an intention. There was a clear path to follow, and that structure supported your progress. It allowed you to observe more carefully, to act more deliberately, and to remain consistent even when motivation fluctuated.

But what comes next is different.

There will be no daily prompts waiting for you. No fixed progression guiding each step. No external system telling you when to reflect, when to act, or how to adjust.

This is where the work becomes yours.

Not something you follow, but something you live.

From Structure to Practice

Over these ninety days, structure served its purpose. It helped you slow down. It helped you notice. It helped you build consistency where there may once have been hesitation or distraction.

But structure was never the goal.

The goal was independence—the ability to think clearly without needing constant guidance, to act with intention without instruction, and to adjust without waiting for direction.

This is the shift from learning to living.

Maintaining Awareness

Everything begins with awareness, and everything returns to it.

Without it, patterns return quietly. Old habits reappear. Reactions begin to feel automatic again, and direction becomes less clear.

But awareness interrupts that process.

It creates a space—small, often unnoticed—between what happens and how you respond. And in that space, there is always a choice.

This space remains available to you at all times.

The practice is simple, but it requires consistency: ***observing what is happening, noticing your response, and adjusting when necessary.***

Stability Over Intensity

Progress is rarely built through extremes.

It is built through consistency—through small actions repeated over time.

There will be days when your focus is strong, and your energy is steady. There will also be days when it is not. Neither defines your direction.

What matters is that you continue.

Avoid the tendency to overcorrect. Avoid the need to do everything at once. Stay steady, even when progress feels slow.

Consistency creates stability. And stability sustains growth.

Responding to Disruption
Disruption will come.

There will be moments when your attention drifts, when structure fades, when consistency weakens. There will be unexpected situations, periods of fatigue, and moments where progress feels less visible.

This is not failure.
It is part of the process.
The difference now is your response.

You recognize it sooner. You pause earlier. You return with less resistance. What once might have taken days now takes moments.

Do not try to eliminate disruption. Develop the ability to move through it and return to alignment.

Refining Your Approach
You no longer need a fixed system, but you may still choose to use one.

If something supports your clarity, keep it. If something strengthens your consistency, refine it. Journaling, reflection, structured habits—these can still serve you, but only if they remain useful.

Remove what is unnecessary. Keep what is effective.

Your structure should support your life—not control it.

Applying in Real Conditions

Practice only becomes meaningful when it is applied in real situations.

Not when conditions are ideal, but when they are not.

When time is limited. When pressure is present. When emotions are active, clarity feels less accessible.

This is where your progress becomes visible—not in isolation, but in interaction.

In these moments, return to what you have practiced: pause, observe, and respond with intention.

Long-Term Direction

There is no final version of you to arrive at.

There is only continued adjustment.

Your direction will evolve. Your priorities will shift. Your circumstances will change. This is not something to control—it is something to work with.

Your role is not to fix everything permanently.

Your role is to remain aware and adjust when needed.

Clarity does not come from controlling everything around you. It comes from staying aligned within it.

The Process Behind the System

What shaped this process was not theory, but experience.

Recovery was not a single moment. It unfolded over time, through repetition and adjustment.

There were days when progress felt steady, and others when the body required more rest than expected. Sleep became essential. Without it, the body responded with discomfort. With it, stability returned.

Returning to work followed the same pattern. It did not happen all at once. It began gradually—one hour, then two, then three. Each step was based not on expectation but on capacity.

This required awareness.

There was no advantage in rushing. No benefit in ignoring signals. Progress came from respecting limits while slowly expanding them.

At the same time, the structure remained.

Training continued. Writing continued. Reflection continued. Even when the pace slowed, the direction did not change.

This created stability.

Not because the conditions were ideal, but because the approach remained consistent.

Over time, the body adapted. The mind adjusted. What began as recovery became a way of operating.

Not temporary.
Repeatable.

If this practice has meant something to you, it does not stand alone. The story behind it—the accident and the recovery—lives in my memoir, *The Magic of Gratitude*. The philosophy beneath it is explored in *Thinking Like a Stoic*. And its deepest questions—why we suffer, and how the soul survives it—I explore in *Job: The Soul of Stoic Wisdom*. This book is where those ideas become daily action. Those others are where they become understanding.

Final Reflection

What practices from these ninety days will you continue—not occasionally, but consistently?

How will you maintain awareness without relying on external structure?

When disruption appears, how do you now respond—and how quickly do you return?

And with greater clarity than before, what direction are you choosing to continue toward?

Afterword

Freedom is not found in ease, but in disciplined response.

What you have completed is not the end of something.

It is the beginning of how you choose to live.

These ninety days were never meant to change you temporarily. They were designed to show you what becomes possible when attention, discipline, and action are aligned over time—not in theory, but in practice.

There will be no perfect continuation from here. Some days will feel clear, others will not. There will be moments of strength, and moments where old patterns return without warning.

This is not failure. It is part of the process.

What matters now is awareness.

You have learned to observe before reacting, to pause rather than rush, and to act with intention rather than on habit. These are not temporary skills. They are foundations—quiet, but reliable.

Life will continue to present situations that cannot be controlled. Pressure, uncertainty, and discomfort will not disappear. But something within you has changed.

You are no longer dependent on ideal conditions to function well.

Clarity does not come from controlling everything around you. It comes from maintaining stability within yourself.

And this is where the real work begins—not because it becomes harder, but because it becomes yours.

There is no longer a structure guiding each day. No progression to follow. No external system holding you accountable.

Only awareness, choice, and action.

And that is enough.

Over time, you will return to this process—not always deliberately, but naturally. You will recognize when you are drifting. You will pause sooner. You will adjust with less resistance. What once required effort will become part of how you move.

Consistency will matter more than intensity.

Small actions, repeated over time, will continue to shape your direction.

There is no final version of you to arrive at.

Only a direction to maintain.

You do not need to become someone else. You need to remain aligned with what you have already seen and understood.

That requires discipline. It requires honesty. And above all, it requires loyalty to yourself.

Because no system, no philosophy, and no method can replace that.

What remains is simple, but not always easy:
You are responsible for how you think, how you act, and how you respond.

And in that responsibility, there is freedom.

Appendix A

Daily Practice Framework

The purpose of this framework is to provide a simple structure that can be used beyond the ninety-day system.

It is not something that must be followed exactly. It is a reference—something to return to when clarity is needed, and something to adjust as your awareness develops.

Used consistently, even in its simplest form, it helps maintain direction.

Morning Alignment (5–15 minutes)

Begin the day with awareness, before external input has influence.

Sit in stillness for a few moments and observe your thoughts without reacting to them. Allow the mind to settle naturally, without force. From that state, set a clear and deliberate intention for the day.

You may ask yourself:
What matters most today?

How will I respond under pressure?
What kind of person will I be in my actions and decisions?

This is not about controlling the day, but about establishing direction before it begins.

Midday Reset (1–5 minutes)
At some point during the day, pause briefly to restore clarity.

Step away, even for a moment. Take a few controlled breaths and observe your current state—your thoughts, your focus, and your level of presence.

Ask yourself:
Am I acting with intention, or reacting automatically?
Where has my attention been, and where should it return?

If needed, make a small adjustment. A single moment of awareness can shift the direction of the rest of the day.

Evening Reflection (5–10 minutes)
End the day with reflection, not judgment.

Take a few minutes to review your actions and responses with honesty and without criticism. The purpose is not to evaluate yourself harshly, but to understand clearly.

What did I handle well today?
Where did I react instead of respond?
What can be adjusted tomorrow?

Keep it simple. Depth is not required. Consistency is.

Core Principle
Awareness leads to action.

Action reveals patterns.
Reflection creates understanding.
Understanding allows adjustment.
And then the process continues.

Appendix B

Beverages for Clarity, Energy, and Recovery

These practices became part of my routine during recovery, when simplicity and consistency were essential. I continue to prepare these drinks regularly, often every other week, as part of maintaining balance and stability.

The discipline of the mind is supported by the condition of the body. What you consume influences clarity, energy, and recovery—not through intensity, but through consistency over time. These practices are simple, repeatable, and designed to support stability rather than excess.

Warming Infusion (Circulation Support)

Ingredients

- Rosemary
- Turmeric
- Cinnamon

Method
Boil the ingredients in water for 8–10 minutes. Strain and drink while warm.

Use
3–4 times per week, especially when the body feels low in energy or requires warmth.

Morning Clarity Blend

Ingredients
- Carrot
- Green apple
- Lemon
- Ginger

Method
Blend thoroughly and drink fresh.

Use
Best taken in the morning as part of a consistent routine.

Daily Reset Blend

Ingredients
- Beetroot
- Carrot
- Kiwi
- Water

Method
Blend and consume immediately after preparation.

Use
Morning or midday, depending on need.

Endurance Blend

Ingredients

- Beetroot
- Celery
- Lemon
- Ginger
- Green apple

Method

Blend thoroughly and drink on an empty stomach.

Use

When sustained energy or recovery support is needed.

Closing Note

Keep intake simple. Avoid excess.

Support the body so it can support the mind.

What is done consistently will have a greater impact than what is done occasionally.

Appendix C

Personal Tracking & Reflection

This section is designed to be used repeatedly.

It is not tied to the ninety-day structure, but can be applied at any time to maintain awareness, direction, and consistency. Use it as a reference—return to it when clarity is needed, and adapt it as your practice develops.

Daily Check-In
At the end of the day, take a brief moment to observe without judgment.

- What did I focus on today?
- How did I respond under pressure?
- What pattern appeared again?

Weekly Review
Once a week, step back and look at your progress more broadly.

- What improved this week?
- Where did I lose awareness?
- What needs adjustment?

Monthly Reflection

Over time, patterns become clearer. Use this moment to assess direction.

- What habits have become stable?
- What still requires effort?
- What direction am I moving toward?

Simple Tracking Model

If useful, you may choose to track a small number of consistent variables:

- Sleep consistency
- Physical movement
- Focus periods
- Emotional stability
- Daily reflection

Only track what you are willing to maintain. Simplicity increases consistency.

Principle

Clarity is not created through complexity.
It is maintained through consistency.

Quotes & Reflections

These are not meant to be read quickly.

They are meant to be returned to—quietly, repeatedly—when clarity is needed, when direction feels uncertain, or when the mind begins to drift.

Timeless Foundations

"You have power over your mind, not outside events.
Realize this, and you will find strength."
— Marcus Aurelius

"No man is free who is not master of himself."
— Epictetus

"The impediment to action advances action. What stands in the way becomes the way."
— Marcus Aurelius

"He who conquers himself is the mightiest warrior."

106 Quotes & Reflections

— Confucius

"It is not what happens to you, but how you react to it that matters."
— Epictetus

"Small disciplines repeated with consistency every day lead to great achievements gained slowly over time."
—John C. Maxwell

"A positive mental attitude is the right mental attitude in all circumstances."
— Napoleon Hill

"The greatest victory is the one over oneself."
— Aristotle

"Do not conform to the pattern of this world, but be transformed by the renewing of your mind."
— Romans 12:2

"Happy is the one who keeps on enduring trial."
—James 1:12

Reflections

Clarity is not found in controlling everything around you. It is found in understanding what is within your influence—and in acting on it consistently.

Discipline is not restriction. It is alignment. It is the quiet decision to return, again and again, to what strengthens you.

What you repeat shapes your direction. Whether intentional or not, your patterns become your path.

Awareness creates space. And within that space, there is choice. What you choose—consistently—becomes how you live.

You do not need more time. You need more presence within the time you already have.

Stability is built quietly, in moments that seem small as they happen. Over time, those moments become structure.

What is avoided does not disappear. It remains, influencing from a distance. What is observed can be understood. What is understood can be changed.

You are not defined by your first reaction. You are defined by what you choose after it.

Gratitude shifts attention. It does not remove difficulty, but it changes how you meet it.

Faith is not passive. It is the steady decision to continue forward, even when clarity is incomplete.

Prayers & Inner Alignment

These are moments of stillness, intention, and alignment—expressed in different forms, but grounded in the same purpose: clarity, gratitude, discipline, and direction.

Morning Alignment

Let this day begin with clarity and gratitude.
Help me act with discipline, respond with patience, and remain steady regardless of circumstance.
Guide my thoughts toward what is useful, and my actions toward what is right.

108 Quotes & Reflections

Strength in Difficulty

When pressure arises, let me remain grounded.
Help me respond with awareness instead of reaction.
Let difficulty refine me, not define me.
Give me the strength to continue forward, even when the path is unclear.

Gratitude in the Present

Thank you for this moment—for the ability to think, to act, and to choose.
Let me not overlook what is already present.
Help me remain aware of what is good, even within what is difficult.

Evening Reflection

What was done today is complete.
What remains can be addressed tomorrow.
Let the mind settle.
Let the body rest.
Let clarity return on the next day.

Faith & Direction

Help me trust what I cannot fully see.
Guide my direction when certainty is absent.
Let my actions remain aligned with truth, even when it requires discipline.

Closing Reflection

Return to stillness when needed.
Return to awareness when distracted.
Return to disciplined action when it is time to move forward.
And continue—with clarity, with intention, and with quiet strength.

Acknowledgments

This work was not created in isolation.

It is shaped by experience, by challenge, and by the many people who, directly and indirectly, influenced the path that led here. No meaningful process unfolds alone, even when much of it is carried out in silence.

I would like to express my sincere appreciation to those who offered support during difficult periods—through presence, conversation, and quiet acts of care. These moments, though often simple, carried weight. They made the process steadier, and in many ways, possible.

I am also grateful for the individuals I work with each day. Supporting adolescents facing different mental and developmental challenges continues to deepen my understanding of human behavior, resilience, and the importance of patience, structure, and awareness. The lessons found in that work extend far beyond any single setting.

I am deeply grateful to Jehovah for the gift of life, for protection, and for the strength to continue forward with clarity and purpose. In moments where direction was uncertain, that foundation remained.

I also acknowledge the thinkers whose ideas have endured across time—within Stoicism, psychology, and spiritual traditions. Their insights continue to guide, challenge, and refine the way we understand the mind, behavior, and meaning.

And finally, to you—the reader:

Thank you for your time, your attention, and your willingness to reflect.

The value of this work is not found in what is written here, but in what is applied beyond it, in the quiet decisions made when no one else is watching.

About the Author

Jay Pacheco is a father, mental health worker in Sandefjord Kommune, Norway, author, and the founder of PMA Science LLC. His work integrates Stoicism, psychology, and Positive Mental Attitude (PMA) into practical systems designed to develop clarity, discipline, and mental resilience in everyday life.

Through PMA Science LLC and PMA Science Stoic, he continues to develop practical frameworks that connect ancient disciplines with modern psychology and lived recovery.

Following a near-fatal car accident in 2024, he applied these principles directly in his own recovery—physically, mentally, and emotionally. What began as a period of uncertainty became a process of testing, refining, and living the methods he now shares, shaping them through experience rather than theory.

Alongside his writing, he works with adolescents facing a range of developmental and psychological challenges, applying structured approaches to behavior, awareness, and emotional regulation in real-world settings. This work continues to deepen his understanding of resilience, adaptation, and the practical application of psychological principles.

His approach is grounded in one central idea: that meaningful change is not created through intensity or short bursts of motivation, but through consistent, deliberate action over time.

He does not write from theory alone, but from practice—applying, refining, and continuing the same principles he teaches.

www.ingramcontent.com/pod-product-compliance
Lightning Source LLC
LaVergne TN
LVHW010721061126
851857LV00014B/1286